the complete book of Natural Housekeeping

the complete book of Natural Housekeeping

Casey Kellar

Lark Books
A Division of Sterling Publishing Co., Inc.
New York

Art Director: Celia Naranjo
Photographer: Evan Bracken, Light Reflections
Editor: Dawn Cusick
Production Assistance: Hannes Charen
Editorial Assistance: Heather S. Smith

Library of Congress Cataloging-in-Publication Data
Kellar, Casey.
    The Complete Book of Natural Housekeeping / Casey kellar.
        p.   cm.
    Includes index.
    ISBN 1-57990-229-4
    1. House cleaning.   2. Natural products.   3. Recipes.   I. Title.
TX324.K44 1998
648' .5—dc21   98-7016
                    CIP

10 9 8 7 6 5 4 3

Published by Lark Books, a division of
Sterling Publishing Co., Inc.
387 Park Avenue South, New York, N.Y. 10016

© 2000, by Casey Kellar

Distributed in Canada by Sterling Publishing,
c/o Canadian Manda Group, One Atlantic Ave., Suite 105
Toronto, Ontario, Canada M6K 3E7

Distributed in Australia by Capricorn Link (Australia) Pty Ltd., P.O.
Box 6651, Baulkham Hills, Business Centre
NSW 2153, Australia

Distributed in the U.K. by:
Guild of Master Craftsman Publications Ltd.
Castle Place 166 High Street, Lewes, East Sussex, England, BN7 1XU
Tel: (+ 44) 1273 477374 Fax: (+ 44) 1273 478606
Email: pubs@thegmcgroup.com, Web: www.gmcpublications.com

If you have questions or comments about this book, please contact:
Lark Books, 50 College St., Asheville, NC 28801, (828) 253-0467
Printed in China

ISBN 1-57990-229-4

Acknowledgements

Special thanks are extended to friend Carol Gentry who always "drops every-thing" to help me prepare formulas for photo shoots; to friend Jacki Elsom for listening to problems with sincere interest; and to my husband and staff members who covered for me at work so that I can take time to write my books.

Thanks are also extended to the following people and businesses for assisting with the photography for this book:

Golda and Dane Barker
Brinda Caldwell
Tracy Munn
Dale and Jimmy White and Sukie and Katie
Complements to the Chef
Deal Motor Cars
The Natural Home
all of Asheville, North Carolina

Dedication

This book is dedicated to my Moms.
To my mom, Irene, for generously making her beach house available as a place where I
can write and create in solitude; and to my mom-in-law May, who let me go through
her stack of newspaper clippings and extension notes that she diligently collected over
the years to give me additional ideas for this book.

A special thank-you to Janice Kaplan Osthus for the opportunity she presented
for me to do on-the-spot compounding of cleaning formulas during our adventure
in Europe; thanks also to the team that went with us for their feedback.
This adventure generated the idea for this book.

This book's message to my kids: Keep learning and growing.
Be fearless in the face of change and challenge.

CONTENTS

Introduction

I love research. How any type of compound is put together fascinates me. When I go personal shopping to grocery stores, drug stores, and hardware stores, I always read the labels and look at how the compounds are put together. Often, I comment on the ingredients and break down the whys and whats of the formulations (usually out loud, while my kids protest and my husband smiles politely). I think my label-reading hobby is a natural for me given that I am a professional formulator of naturally based bath, beauty, and home fragrance products.

Several years ago my husband and I were asked to be part of a team of guest instructors for a special event in Germany. After a long flight, our group went to preview the location for our classes and seminars. The place was a mess—not at all suitable for the quality of seminars and classes that were to be held the next day. After a moment of silent panic, we rolled up our sleeves, found brooms, mops and rags, and started to work. We soon realized another big problem: no cleansers. Luckily, my seminars required essential oils (for an aromatherapy and perfumery class) and some base ingredients for soaps, oils, and gels, so I unpacked my supplies and quickly went to work making degreasers, floor cleaners, and polishes. My fellow team members remarked on how well these products were working, and I felt good to be able to provide at a moment's notice.

Making your own cleaning products at home can be a matter of very simple chemistry. Many of the ingredients are already in your home, while a few others are inexpensive and easy to find at your local grocery store or pharmacy. Most of the formulas I have given you are natural and nontoxic. For tough, out-of-the-ordinary cleaning problems, I have also provided stronger formulas. Although these stronger formulas are not as natural, at least the choice to use them is yours and you will know what you are using.

My experience in Germany, along with years of questions from friends and family, gave me the idea for this book. My idea was a home care book that would help people learn to make safe and less expensive products for their home environments. In addition to chapters for your home, I've also included formulas to help you clean home office equipment, vehicles, and even your pets. A special chapter featuring gifts for special occasions is also included.

I hope you enjoy reading and using this book as much as I enjoyed writing it. I know my first copy will go to my daughter as the perfect housewarming gift for her very first home!

Cleaning Basics

A Clean History

We have come a long way from the cave and hut homes of early humans. First came the introduction of soap as a method of cleanliness to defend ourselves against disease; soon we realized that we also needed to keep our food clean and free of germs. Our huts became more than basic shelter—they became "homes."

In the Victorian era we started to collect beautiful things. Formal patterns of entertaining were common, which meant cleaning and maintaining households. During this time stores offered limited cleaning items, so several simple compounds were used to clean many things. Most cleansers were organically based and were made at home.

As time continued, women became "homemakers." The numbers and types of cleansers, polishes, and other "homemaker cleaning" products increased on the shelves. These products were romanced with emotional advertising suggesting that a woman's family would appreciate her much more because she used a certain brand or that her troubles would "go down the drain" if she just used a certain product.

Then came the influx of baby boomers entering the work force, and an increasing percentage of the female population becoming "working mothers." This generation created a demand for fast and easy cleaning solutions with minimal personal and environmental hazards from ingredients. In came the slogans of "Natural Is Better." Consumers expressed concerns about what's in cleaning products, how ingredients are tested, and with recycling. Many of us now find we are spending more and more time "cocooning" in our homes. Home offices, Internet shopping, and other conveniences have changed the way we view our homes. A home today is more than a place to eat and sleep. Now a home is a livable space for working, entertaining, relaxing, and everyday living.

This book will help you understand how to make both natural and simple industrial compounds for every occasion and reason at home. There are solutions for your home, office, garden, garage toys, and pets. We have attempted to take the "industrial" out of the cleaning solutions as much as possible and to create natural beauty products for your home.

Materials and Supplies

Containers

If you are going to be packaging your own cleaning supplies, you will need to start saving special containers. There are many containers you get daily with your regular grocery shopping that are suitable for storing your new cleaning solutions. Milk jugs, syrup bottles, ketchup bottles, wine bottles, and jelly jars all work well, but the possibilities are almost endless. If you have something prettier in mind, shop at garage sales for older, decorative bottles you can use for yourself or for gift giving. Keep a collection of these as you find them and then they will always be available for you to "whip" something up for a great gift.

When preparing containers for storing cleaning formulas, be sure to follow the following tips. First, thoroughly clean every container and rinse well before using. Next, always label the container with the date, the contents, and the expected shelf life. (We have included approxi-mate shelf life dates at the end of each recipe.) Finally, take great care to find a cool, dry storage place away from the reach of small children and pets.

Scrub Brushes, Sponges, and Rags

Collect an array of cleaning aids and keep them in a box so that they are easy to find. A friend of mine who is a designer by profession keeps a pretty box in her laundry room filled with brushes, bright-colored, rolled up rags, sponges, and feather dusters. The contents of the box are arranged so delightfully that you think you're in a gorgeous guest bedroom for a moment when you walk into her laundry room. Whether you choose to go all out or not, at the very least you should assemble the following items and keep them in a handy place.

Baby Bottle Brushes: These brushes are great for cleaning specialty bottles with hard-to-reach places).

A Variety of Sponges (man made and natural): The best natural sponges have a darkish tinge to them because they haven't been bleached. ((The bleaching process weakens the sponge, ruining their effectiveness for heavy-duty jobs.) If you are evnironmentally concerned about the way sponges are harvested, be sure to ask the store owner how the sponge was harvested. Harvesting with the dragging method can damage ocean reefs, whereas cutting the sponges off by hand allows reef sponges to regenerate.

Cotton Rags: Make your own beautiful rags by recycling cotton t-shirts and underwear that have outlived their usefulness as clothing. (Chamois for car cleaning is the only commercial "rag" I buy.) Just be careful to test colored rags with your cleaning compound or you will end up painting instead of cleaning.

Two Basic Brushes: A stiff-bristled brush is a cleaning must for tough jobs on tough surfaces. A soft-bristled brush and sponges work best when you're working on a scratchable surface. Remember to always test surfaces!

Ingredients

Many of the ingredients used in the cleaning formulas in this book are everyday household items. In fact, they may already be sitting in your kitchen cabinets. These items include baking powder, lemon juice, table salt, onions, garlic cloves, canola oil, beeswax, milk, and black pepper, just to name a few. Other items, such as those listed below, may require a trip to the store.

Ammonia: A commercial chemical found in fertilizers, fuels, cleaning products, and some hair products. Ammonia fumes can be extremely irritating and can react with other household chemicals, such as hydrogen peroxide and other commercial compunds, to produce a lethal gas. **Use caution when using, use only as directed in this book, and always wear gloves.**

Apricot Kernel Oil: A light, natural oil that can be found at health food stores.

Arrowroot Powder: This white powder comes from the starch in the root of the arrowroot plant. (The arrowroot plant was once used by Native Americans to heal wounds caused by poison arrows.) It has properties similar to cornstarch but has a finer consistency—somewhere between cornstarch and rice flour—and superior drying abilities. It can be found in health food stores.

Aspirin (salicylic acid): Pure aspirin is actually a mild acid, which can come in handy for cleaning. Be sure to buy pure aspirin, not one of the analgesic blends. It's available at drug and grocery stores.

Baking Soda (sodium bicarbonate or bicarbonate of soda): Baking soda is used as an odor-removing agent and to raise the pH of a product (which helps stop corrosion). It is also used in baking and can be purchased in any grocery store.

Beer: For use in this book, any brand of cheap, pale beer will do.

Borax: (sodium tetraborate/sodium borate) Borax is mined in dry, arid areas such as deserts. It makes an excellent water softener and is sometimes used in cleaning agents to help penetrate and lift out dirt. It can also be used as an abrasive if used in paste form.

Bran: Bran is the most fibrous part of a variety of whole grains. For recipes in this book, use the plain, natural bran found in the bulk bins at health food stores.

Butter: Butter is used in cleaning recipes as a softening agent or to make the compound glide on. Margarine will also work.

Facing page, clockwise from top left: apricot kernal oil, sodium bisulfate, cedar oil, butter, cornmeal, and aspirin.

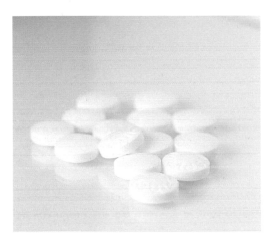

Carnauba Wax: This heavy wax substance is used in some candies and for automobile wax. Look in your hardware store's automotive section.

Castile Soap: Named for the region in Spain where it originated, this very mild soap is made with olive oil. Liquid castile soap is often used as a foaming base and can be found in drug stores or health food stores.

Cedar: Cedar wood shavings and essential oil can be found in garden centers and health stores, respectively.

Chamomile: The flowers of this herb are available in both extract and essential oil forms. Look for them at health food stores.

Chlorine Bleach: A bleach with a chlorine base. (Most bleach is a chlorine bleach.) Use the same precautions as gasoline or turpentine. Do not spill on delicate surfaces in its pure form or an unwanted bleaching may happen.

Comfrey Root: Available from the plant in a crushed form in your health food store. Also available as an essential oil.

Cornstarch: A natural powder that is used as a thickening agent in cooking, body care, and cleansing agents to improve absorption. Cornstarch also helps buffer acids to achieve a more alkaline pH balance. Available in the baking section of grocery stores.

Essential Oils: These pure "essence" oils are formed by extracting fragrant oils from plants through steam or maceration techniques. Essential oils can be found in health stores. Following is a list of the essential oils often used in natural home cleaning: cinnamon oil, orange oil, lemon oil, rosemary oil, fir oil, balsam oil, cedar oil, nettle oil, pennyroyal oil, citronella oil, eucalyptus oil, and chamomile oil.

Ethyl Alcohol: This is the type of alcohol found in alcoholic beverages such as beer, wine, and liqueur. When a recipe in this book calls for ethyl alcohol and does not specify beer or wine, use a clear, high-alcohol form such as vodka.

Fuller's Earth Clay: This natural clay from the earth works in cleaning recipes as a "drawing" agent to draw out stains from porous surfaces. Look for it in health food stores.

Gasoline: Derived from the distillation of petroleum, this liquid hydrocarbon **is combustible and should be stored in a cool place. Take care to avoid the fumes and wear gloves when handling.**

Gelatin: Gelatin works as a thickening agent and is commonly available in grocery stores. Use only plain, unflavored gelatin for the formulas in this book.

Glycerine: Glycerine is a naturally produced emollient derived from vegetable or animal sources or through synthetic production. You can buy liquid glycerine by the bottle in pharmacy or health food stores.

Glycerine Soap: Soap made from glycerine and oils has a lovely, translucent quality. For the recipes in this book, look for soap made from vegetable-derived glycerine.

Facing page, clockwise from top left: chamomile oil, potash, oranges, liquid whiting, assorted essential oils, and comfrey root oil.

Isopropyl Alcohol: Commonly known as rubbing alcohol, isopropyl alcohol is often used as a topical antiseptic and for household sterilization. It is available over the counter at most pharmacies. **Not safe for consumption.**

Liquid Fabric Softener: A product or group of products used to break the surface tension and decrease static in clothes caused by clothes dryers. Most commercial fabric softeners use fragrance for a fresh smell. Available at grocery stores.

Liquid Iron: This is the vitamin iron suspended in a liquid solution. Available in garden centers. **Not for consumption.**

Liquid Whiting: Also known as liquid bluing, this product is used as a bleach, clarifying through a blue bleach process that counteracts the yellow and gray aging of clothing. Liquid whiting can also clarify water if used in small amounts. Look for it in the laundry sections of larger grocery stores. **Take care to avoid the fumes and wear gloves when handling.**

Mild Detergent: Mild detergents in liquid form can be found in the cleaning section of most stores; verify its mildness by searching the ingredients list for sodium laurel sulfate or ammonium laurel sulfate.

Muriatic Acid: This strong, **caustic acid** is used in products designed to clean concrete and tile. **Use the same care as other acids for storage and handling.** Use it only to remove tough rubber, mastic, and mortar from tile, plaster etc. It should always be diluted with water for handling, adding the acid to water and not vice versa. When mixed in a cleaning formula, use immediately and discard any remaining formula by flushing it down the drain with lots of water. Available at most hardware stores.

Petroleum Jelly: This petroleum distillate is often used as a coating agent and lubricant. When making the formulas in this book, use only pure petroleum jelly (available in small containers in retail stores).

Potash: Potash is made from ashes that enrich the soil and is available from your favorite garden center. If you cannot find a commercially-packaged product, you can make your own by cleaning your fireplace. Put the ashes in a sealable plastic bag and seal, then manipulate the bag until the ashes are powder. Add to the formula as "home made" potash.

Propylene Glycol: This is an alcohol-blend ingredient that serves as a solvent for essential oils. It is also used in preparation of herbal extracts in some commercial varieties. Pharmacies and hardware stores can usually order it for you if you can't find it on the shelves.

Red Turkey Oil: Also known as sulfated caster oil, it can be found at health food stores.

Rice Flour: Similar to cornstarch in its characteristics and uses, rice flour is sold as a very fine powder. Just as its name suggests, it comes from rice. Available at health food stores.

Facing page, clockwise from top left: beer, washing soda, apple cider vinegar, vanilla beans, petroleum jelly, and rice flour.

Sand: Choose a fine grind sand for the recipes in this book. If you don't live by a beach, an aquarium store is a good source.

Sour Milk: Sour milk is easily made by adding 1½ tablespoons (22 mL) of lemon juice or 1⅓ tablespoons (20 mL) of vinegar to 1 cup (236 mL) of lukewarm milk. Allow to sit for several minutes before using.

Sodium Bisulfate: This salt-derived acidic product is used to lower the pH of things because of its mild acidic content. Sodium bisulfate is commonly used as a pH adjuster in pools and spas, so it can be found in pool and spa maintenance sections in stores. Alternately, ask your pharmacist to order it for you.

Sugar and Unrefined Sugar: Sugar is available at the grocery store in either a refined state, which is white, or unrefined, which is usually light brown (unbleached) and slightly granular.

Sweet Almond Oil: This natural oil is expressed from the almond nut. Look for it in health food stores.

Tea-Tree Extract: This is a byproduct of the Teatree (known botanically as *Melaleuca alternifolia*) most commonly found in Australia. It is nature's antifungal product and can also work as an antibacterial agent. It is often used cosmetically for its astringent properties. If you smell it in its essential oil or extract form, you will find it has a most pungent and medicinal smell.

Trisodium Phosphate: This phosphate-based product works as a strong, basic cleaning compound. Look for it in feed and hardware stores. If you can't find it in its pure form, ask them to order it or read the ingredients on some common cleansers. (Some cleansers are mostly made up of this one ingredient.)

Turpentine: This very strong solvent is extracted from pine oil. It is flammable in both the liquid and fume form, so use extra caution. Use gloves when handling and avoid inhaling the fumes.

Vanilla Bean: Whole vanilla beans at your local health food store.

Vinegar (apple cider and white): Vinegar is made through a fermentation process using fruit or wine. It is commonly used for cooking purposes, but it is also a wonderful natural degreaser. Look for it in the grocery store.

Washing Soda (sodium carbonate): Also known as soda ash, this naturally-occurring mineral is found in sea beds. It is used in soaps and other cleaning compounds, as well as antacids. Look for it in the laundry section or in a health food store.

Wheatgerm Oil: This natural oil is made from wheat germ. High in vitamins A, D, and E, this oil also contains lecithin. It has a longer shelf life than some oils, and can be found in health food stores.

Yogurt: When referred to in this book for cleaning recipes, use the plain yogurt without fruit, color, or flavorings.

Facing page, clockwise from top left: sugar, lemons, yogurt, sour milk, yogurt, trisodium phosphate, and almond oil.

General Living Areas

General living areas contain a challenging assortment of materials that need cleaning. Carpet, tile, vinyl, and wood floors all take the brunt of daily foot traffic. Walls, though often ignored, also accumulate dirt and grime, and can add a dingy feel to an otherwise clean room. And though we often take its comforting presence for granted, our furniture also needs cleaning care.

In this chapter we will cover the living areas of your home and your treasured decor. It comes as no surprise that our homes are a reflection of our interests and our personalities. Look around you at your main living space. Perhaps you have hired a professional decorator to design and arrange your treasures. Maybe you simply have years of memories and possessions you love that you have arranged yourself. Or maybe

you're just starting out with a few hand-me-downs and garage sale items to fill in.

If you are fall into one of the first two categories, you probably have a substantial investment (financially, emotionally, or both) in your possessions, and you need to keep them looking beautiful without putting yourself or your home at environmental risk. If you are just starting out, some of your new "used" items may need cleaning up prior to using.

Carpets

Wall-to-wall carpeting and decorator rugs add comfort and style to the rooms they grace, but they do require consistent care to maintain thir beauty.

General Carpet Cleaner

- ⅓ cup (79 mL) soap flakes (See recipe on page 64.)
- 3½ cups (829 mL) boiling water
- 2 teaspoons (10 mL) of washing soda
- 6 drops vanilla essential oil (cuts urine and smoke odors)
- 6 drops peppermint or wintergreen essential oil (mild disinfectant)

Stir the soap flakes into the boiling water until they dissolve. Let cool to room temperature, then add washing soda and essential oil. Mix and put in glass jar or bowl. Spot test a small, inconspicuous area of your rug first to test for color stability of the fibers. To use, shake well and gently scrub into carpet with a brush, then let dry and vacuum. Avoid saturating the carpet to prevent stretching. Shelf life: approximately 6 months.

Dry Carpet Cleaner and Deodorizer

- ¾ cup (177 mL) baking soda
- ¼ cup (59 mL) cornstarch
- 5-6 drops of your favorite essential oil

Spot test a small, inconspicuous area of your rug first to test for color stability of the fibers. Put the baking soda, cornstarch, and essential oil through a sifter until blended. Sprinkle lightly on your carpeting and allow to sit several hours or overnight. Vacuum up and enjoy your fresh, fragrant carpet.

Carpet Stain Removers

Since carpets are usually a fabric, refer to the extensive spot cleaning section beginning on page 70.

Carpet Revitalizer

- 1 cup (237 mL) white vinegar
- 3 cups (711 mL) warm water
- 2 cups (474 mL) baking soda
- 2 tablespoons (30 mL) cornstarch
- 4 drops vanilla essential oil

This is a two step process. First, mix the vinegar and water together and lightly scrub with a brush (do not saturate carpet). Let dry to "release" grime from the surface area. (Spot test a small, inconspicuous area of your rug first to test for color stability of the fibers.) Store excess in a glass jar. Shelf life: 3-4 months.

Next, sift the baking soda, cornstarch, and essential oil together using a hand sifter. (You may need to do this a few times because the essential oil will clump at first.) Sprinkle over dry carpet and let sit for an hour, then vacuum. The powder will "lift" any oils and additional grime that has floated to the surface during the first process and will finish the deodorizing process. Powder shelf life: discard after 2 weeks.

Wood Floors

If your wood floor is in extremely bad shape, you may have to strip all the old wax off, do a light sanding on the worst parts, and then "rewax" your floor with a wax paste.

Rub-a-Dub Scrub for Wood Floors in Good Condition

- 2 cups (474 mL) soap flakes (See recipe on page 64.)
- 1 cup (237 mL) Fuller's earth clay
- ½ cup (118 mL) cornstarch
- 1½ cups (355 mL) washing soda
- 5 cups (1.18 L) water

Mix the ingredients together in a large pan and bring to boil while stirring. Lower the heat to simmer and stir for another 10 minutes. Cool and store in a jar. Apply small amounts with water to the floor using a handled sponge mop, wringing out regularly to remove dirt from sponge. Rinse and let dry. Shelf life: approximately 2 months.

Wood Floor Polish

- ½ cup (118 mL) beeswax
- ¼ cup (59 mL) linseed oil
- 2 tablespoons (30 mL) wheat germ oil
- 1 cup (237 mL) turpentine

Melt beeswax on low heat in pan on stove. When melted add the oils. Remove from heat and stir until cool. Add turpentine, stirring until well mixed. Use sponge to coat the floor and allow to sit 12-24 hours. Polish by hand or with a floor polisher and buff well. Store in a glass jar in a cool area. Shelf life: approximately 4-5 months.

Hardwood Floor Pet Odor Remover

- ½ cup (118 mL) water
- ½ cup white vinegar

Mix together and let sit on floor for ½ to one hour. Using water, rinse and mop up. Be careful with wax-finished floors as the vinegar will remove some of the wax.

If you are fighting pet odors, however, the pet's urine has probably already damaged the finish and you'll have to rewax anyway.

Dry Wood Floor Revitalizer

- ½ cup (118 mL) safflower cooking oil
- ¼ cup (59 mL) red turkey oil (sulfated castor oil)
- 6 drops Canadian balsam fir essential oil

Stir ingredients together and wipe on floor. Let dry. This is a great revitalizer for any wood surface that has been mistreated and dried out. Porous wood will quickly "absorb" this mix and start to look alive again.

30

Tile and Vinyl Floors

If cleaning the floors gives you a Cinderella complex, try these homemade helpers that are sure to speed you up and get you to the party on time!

Handy-Dandy Pine Cleaner

- 1 cup (237 mL) liquid soap blend (See page 64.)
- ½ cup (118 mL) pine oil or extract
- 6 cups (1.4 L) warm water

Mix all ingredients together and store in a plastic jug. Spritz on and wipe or wash off. Store in a cool, dry place. Keep out of the reach of children. Shelf life: approximately 6-9 months.

Lemon Fresh Cleaner

- 1 cup (237 mL) liquid soap blend
- ¼ cup (59 mL) fresh squeezed lemon juice
- ¼ eyedropper of tea-tree extract
- 6 cups (1.4 L) warm water

Mix the ingredients together and shake well. To use, put in a plastic spray bottle, spritz on, and wipe off. For larger jobs, put in a large bucket, with sponge for large jobs. Spritz on and wipe off, or sponge on rinse sponge then use water to wash off. Store in a cool dark place out of the reach of children. Shelf life: approximately 9-10 months.

Heavy Cleaning Solution

- ½ cup (118 mL) sudsing ammonia
- ½ cup liquid soap blend (See page 64.)
- ¼ cup (59 mL) white vinegar
- 6 cups (1.4 L) of warm water

Mix the ingredients together and store in a plastic jug. Spritz on and wipe or rinse off. Store in a cool, dry cool place. Keep out of the reach of children. Shelf life: approximately 4-6 months.

Windows

Few cleaning projects reward you with such glowing results as window cleaning.

Gentle Window Cleaner (for lightweight jobs)

- ½ cup (118 mL) isopropyl alcohol
- ½ cup vinegar
- 2 teaspoons (10 mL) liquid soap blend (See recipe on page 64.)
- 6 cups (1.4 L) warm water

Mix the ingredients together in a large plastic jug and shake well. To use the cleaner, pour some into a plastic spray bottle. Clean windows and mirrors with a soft cloth or newspapers. Shake well prior to each use and store in a cool, dark place. Keep out of the reach of children. Shelf life: approximately 4-6 months.

Non-Streak Window Clarifier

- 1 cup (237 mL) white vinegar
- 2 tablespoons (30 mL) isopropyl alcohol

This recipe helps remove stubborn streaks. Mix the ingredients together and use as you would the Gentle Window Cleaner above. Shelf life: 6 to 9 months.

Paint Remover

- Ammonia
- Water

For old paint that won't scrape off with a razor blade, mix equal amounts of ammonia and water. Saturate the paint on the glass, taking care not to get the solution anywhere else. Allow the solution sit for 3-4 minutes. Peel off the paint with a razor blade, then wash the windows with one of the cleaning solutions above.

Walls

Walls can really be a drag to clean, all that volume of space that we look at every day and so seldom clean. Some of us try to cover as much of that space as possible, but it's still there and needs to be cleaned now and again. Here are some home cleaners that are sure to make that job a snap!

Light Job Cleaner for Painted Walls & Washable Wallpaper

- 1 cup (236 mL) apple cider vinegar
- 1 gallon water

Mix the ingredients together and use to wipe down walls. This recipe also works on vinyl or fiber-coated panels. Shelf life: approximately 6 months.

Heavy Job Cleaner for Painted Walls & Washable Wallpaper

- 3 tablespoons (45 mL) trisodium phosphate
- 1 (3.79 L) gallon of water

Mix the solution with a spoon and wear rubber gloves while scrubbing. If you are cleaning washable wallpaper, make sure not to "over wet" the surface to prevent the wallpaper's glue from dissolving. This recipe also works on vinyl or fiber-coated panels.

Wood Wall Cleaners

- Choose an appropriate recipe from the wood floor or furniture section.

Helpful Hint:

Homemade "hole filler" is easy to make to fill small nail holes prior to painting. Just mix together ¾ part baking soda and ¼ part white household glue mix and use right away. You may have to go back and fill a second time as the mixture will shrink some as it sets. You are now ready for painting or papering!

Furniture

I don't know about you, but I love furniture, and I collect and refinish old antique furniture occasionally as a hobby. Whether your passion is antique, funky eclectic, art deco, Oriental, specialty European imports, family heirloom, or contemporary, I think polishing and cleaning the furniture is fun. Cleaning gives me time to spend a few moments really looking at and appreciating special pieces. Take a moment to clean and polish your furniture, and it will love you back with a "like new" shine.

Basic Furniture Polish

- ½ cup (118 mL) sweet almond oil
- 1 tablespoon (15 mL) wheat germ oil
- 1 tablespoon melted beeswax
- 2 tablespoons (30 mL) red turkey oil
- Several drops of essential oil (optional)

Mix all of the ingredients together to create a great moisturizing polish for dry wood that does not have a heavy wax finish. (If your furniture has a heavy wax or polypropylene finish, then use the Lite Polish below.) Choose one of the variations below to customize your polish. For a lemon oil polish, add 3-5 drops of lemon essential oil.

For a balsam oil polish, add 3-5 drops of fir or balsam essential oil. For a "rejuvenating" polish, add 3-5 drops of cedar oil. For a rosemary/juniper oil polish, add 2 drops of each essential oil. Other essential oils can be used to create any fragrance you like, but be sure not to add more than 5-6 drops because the essential oils are strong enough to work as a solvent and could dissolve your wood finish if used in excess.

Gentle Wood Cleaner

- ½ cup (118 mL) canola oil
- ¼ cup (59 mL) liquid soap blend
  (See page 64 for recipe.)
- ¼ cup water

A good wood furniture cleaner, such as the one above, helps remove dust while restoring shine. Never use straight water for cleaning wood furniture or harsh substances (such as vinegar or ammonia) that will stain your finish. To make this recipe, combine the ingredients and shake well right before using. Apply with a cloth rag a little at a time, then finish with a dry cloth and buff dry. Follow with furniture polish.

Water and Alcohol Wood Stain Remover

- 1½ tablespoon (22.5 mL) turpentine
- ¼ cup (59 mL) boiled linseed oil

This recipe works great for furniture that is not ready for complete refinishing but is dull, water-spotted, or needs serious buffing. First mix the ingredients together and shake well. Apply a small amount to a cloth rag, remove excess, and buff thoroughly. Follow with polish.

Lite Polish for Heavily Waxed or Polished Wood Surfaces

- ½ cup (118 mL) linseed oil
- ¼ cup (59 mL) white vinegar
- Essential oil (optional)

Mix together and shake prior to using. Add 2-3 drops of an essential oil if desired. Shelf life: approximately 6 months.

Furniture Dusting Spray for Wood Furniture

- ½ cup (118 mL) apricot kernel oil
- 1 tablespoon (15 mL) isopropyl alcohol
- 3 tablespoons (45 mL) of liquid soap
  (See page 64 for recipe.)
- 1 cup (237 mL) distilled water

Mix together all ingredients. Shake well, then spray on and wipe off.

Wicker/Rattan Cleaner

- 1 cup (237 mL) hot water
- 2 tablespoons (30 mL) salt
- 2 tablespoons (30 mL) liquid soap blend
  (See page 64 for recipe.)

Dissolve salt in the hot water, then add liquid soap. Mix until all is dissolved and allow to cool. To use, spray on and wipe off. The salt helps soften the wicker or rattan, allowing the dirt to soak away easier. If your wicker or rattan is painted, check it first for color fastness of paint. If the furniture isn't color-fast, just use a soap and water combo.

Plastic/Vinyl Furniture Cleaner

- 1 cup (237 mL) water
- 1 tablespoon (15 mL) vinegar
- 2 tablespoons liquid soap blend
  (See page 64 for recipe.)

This recipe works well to loosen dirt. Mix the ingredients, rub on, and rinse.

Gentle Leather Cleaner

- 1 cup (237 mL) water
- 1 tablespoon (15 mL) liquid soap
  (See page 64 for recipe.)
- ½ cup (118 mL) white vinegar

Mix together and shake before using. Apply to a soft cloth rag and wipe down. Follow with furniture polish. Shelf life: approximately 6 months.

Leather Polish

- ½ cup (118 mL) wheatgerm oil
- ¼ cup (59 mL) castor oil (red turkey oil)

This mixture will soften, condition, and help waterproof your leather furniture. (If your leather is white or very light beige, color test on small spot on back first since this recipe will sometimes slightly darken very light leathers.)

"Dry Clean" for Upholstered Furniture

- ½ cup (118 mL) baking powder
- ½ cup cornstarch

Sift ingredients together and then mix with a small amount of water to make a paste. Apply to the fabric, let sit for 30 minutes, and then brush off with a stiff, bristled brush. Vacuum.

Spray-On Fabric Cleaner

- 1 cup (237 mL) water
- ⅛ cup (29 mL) liquid soap
- ½ teaspoon (2.5 mL) baking soda
- 1 tablespoon (15 mL) white vinegar

Mix the ingredients together with an electric hand mixer. (It will be foamy at first.) Pour into a spray bottle and shake well before using. Spray onto the dirty areas (color test first), then use a damp sponge to lightly scrub. Let dry and vacuum the area.

Spot Fabric Stain Removers

Refer to the extensive spot cleaning section beginning on page 70.

Helpful Hint:

Keep a bottle of club soda and box of baking soda on hand for on the spot staing removal. Use the baking soda to absorb up the excess stain moisture and then brush off quickly to clear the debris and any baking soda left on the spot. Then quickly put some club soda on the same spot. The soda will often "bubble out" the remainder of the stain before it has a chance to set. Leave on one minute, then absorb up with a dry sponge.

Other Challenges

Fireplaces, hearths, and houseplants are frequently overlooked on cleaning days, but clean-up is a snap with these simple recipes.

Brick & Stone Fireplace Cleaner

- ½ cup (118 mL) ammonia
- 1 cup (237 mL) vinegar
- ¼ cup (59 mL) baking soda

Mix the ingredients together. Start cleaning at the bottom of hearth first and work your way up.

Helpful Hint:

The soft drink Coke (original) works as a great stone cleaner because of the acid content. Apply Coke to a stiff bristled brush and scrub. Scrub again with plain water, then blot dry.

Happy Leaves Houseplant Polish

- 1 tablespoon (15 mL) sweet almond oil*
- 4 cups (948 mL) warm water
- 1 teaspoon (5 mL) of unscented gentle shampoo

This recipe cleans plant leaves and gives them a well-conditioned shine. Before applying the recipe, move your plants to the bathtub or outside the back door to avoid unavoidable water messes. To make the recipe, mix all ingredients together in a decorative spray bottle. Because oil and water do not mix, you will need to shake the mixture thoroughly before each use. For delicate leaves, spray on a fine mist and follow with a fine mist of plain water. For large, waxy leaves, spray the mix on and then wipe off. Shelf life: 1 to 2 months.

*Don't be tempted to substitute a mineral oil for the natural oil in this formula as it will coat the leaves and diminish the plant's carbon dioxide and oxygen intake. The natural oil allows the plant to "breathe."

Helpful Holiday Hint:

To keep your Christmas tree looking fresh, mix ¼ cup (59 mL) liquid iron (available in the garden section of hardware stores), 1 cup (237 mL) corn syrup (light), and 2 tablespoons (30 mL) bleach (works as an antifungal). Make a fresh cut to the base of the tree if needed (cuts will "seal" themselves after a while). Add 3-4 tablespoons (45-60 mL) to tree water each time you refresh the water. The bleach will work as an antifungal to keep water from going stale. If you have a small pet who might drink from the base water, just make the formula without the bleach.

Kitchens & Bathrooms

Kitchens and bathrooms are heavy traffic areas, used daily, and they bring lots of cleaning challenges. Some people love cleaning as a form of relaxation; however, even people who hate cleaning will need to clean these areas of the house. In this chapter I will attempt to show you how fun it is to discover what ordinary, well-known ingredients can do. Many simple products can easily do double duty around the house as cleaning products!

The recipes always start with the least caustic and abrasive cleanser and then move on to stronger, more industrial cleaning solutions for heavy-duty problems. Keep in mind that gentler is almost always better. You can sometimes harm the finish if your cleaning solution "overkills" the problem (as some of the more caustic commercial cleansers often do).

Because sanitation is an important aspect of kitchen and bathroom cleaning, we sometimes need to use alcohol or ammonia in some of the cleaning and disinfecting solutions. It is a comfort to know the items you are working with and that you are empowered to control the strengths. Also, these formulas will give you the same "industrial" effectiveness of commercial cleaners while being both safer, milder, and much less expensive. For cleaners and polishes that do not require disinfectant properties, we have stayed with the tried-and-true natural ingredients that we all know and love.

Sinks, Drains, Tubs, and Toilets

Sinks, drains, tubs, and toilets are the worst! This is where the nitty-gritty, heavy cleaning needs to be done. (Floors can qualify for this award too, but more on them later.) This section provides both lightweight, natural-based recipes and some stronger, more industrial-strength recipes for really tough jobs. I really believe in natural first, but mold and bacteria are also natural and aren't nice. Occasionally we need to do battle with a few elements of nature, so roll up your sleeves and let's get busy!

Liquid Tub and Tile Cleaner

- ½ cup (118 mL) ammonia
- ½ cup (59 mL) white vinegar
- ¼ cup (5 mL) baking soda
- 1 teaspoon borax

Mix the ingredients together and place the mixture in a squeeze-top container. Stir or shake before use. Use with scrub brush or toilet brush.

Store in a cool, dry place and keep out of reach of children. Shelf life: approximately 6 months.

Scouring Cleanser for Plastic, Tile, Fiberglass, and Ceramic Surfaces

- ½ cup (118 mL) liquid soap blend (See page 64.)
- 1 teaspoon (5 mL) borax
- 2 teaspoons (10 mL) baking soda
- 1¾ (403 mL) cups very warm water

Pour the water into a stainless steel mixing bowl and add the liquid soap. Stir well, then add the dry ingredients. Stir the mixture until it's grainy. Store in a squeeze top container (old ketchup containers work well) and stir or shake before using. Use with scrub brush. Keep out of the reach of children. Keep the cleaner in a cool, dry place.

Shelf life: approximately 3-4 months. Note: If you have a stubborn stain on a fiberglass or plastic surface, you can add a few drops of peroxide to this mixture and "scrub" away the stain.

Disinfectant

- ¼ cup (59 mL) liquid bleach
- ¼ cup lemon juice
- 3 cups (711 mL) water

Helpful Hint:

Try this alternative if you don't want to handle liquid bleach.

- ¼ cup (59 mL) powdered laundry detergent with bleach
- ⅛ cup (29 mL) water

Mix into a paste and test a small sample area for abrasive scratching. Work the paste onto the mildewed area. Use any leftover paste in your laundry right away.

Mix and use for toilet bowls, tubs, showers, or sinks that need disinfecting. This formula is also great for cutting mold and mildew that has started to creep in tile grout and around showers. Rub on the affected area, let sit for 5 minutes, then rinse off. For smaller projects, I recommend that you cut this formula in half. Leftover disinfectant should be poured down your drain with running water.

Here are some important rules about this disinfectant formula.

1. Keep out of the reach of children.

2. Do not use on fabrics.

3. Use gloves and avoid skin exposure.

4. Never, never mix bleach with ammonia; dangerous fumes can result.

5. Never leave the disinfectant soaking in the toilet bowl without closing the lid and supervising to prevent thirsty pets from being harmed.

Mirrors & Windows

Making your own window and mirror cleaner is a simple and inexpensive proposition. If you haven't tried cleaning windows with newspapers before, you're in for a surprise. Newspapers are an age-old window shining technique. The ink on the paper works like "bluing" works on laundry, shining the windows without leaving streaks. Make sure you use gloves as the print may get on your hands. For streaks that remain, spray on more cleaner and go at them again.

General Window Cleaner

- 3 teaspoons (15 mL) liquid soap blend (See page 64 for recipe.)
- ¾ cup (177 mL) white vinegar
- ½ teaspoon (2.5 mL) baking soda

Mix the ingredients together in a large plastic jug and shake well. To use the cleaner, pour some into a plastic spray bottle. Clean windows and mirrors with a soft cloth or newspapers. Shake well prior to each use and store in a cool, dark place. Keep out of the reach of children. Shelf life: approximately 4-6 months.

Mirror Degreasing Formula

- 3 cups (711 mL) isopropyl alcohol
- 1 cup (237 mL) ammonia
- ½ cup (118 mL) liquid soap blend (See page 64.)
- 10 cups (2.37 L) water

Follow the same mixing and storage directions as the formula above. Shelf life: approximately 6-9 months.

Toothpaste Remover

- Ammonia
- Water

For old toothpaste that won't come off, mix equal amounts of ammonia and water. Saturate the paint on the glass, taking care not to get the solution anywhere else. Allow the solution to sit for 3-4 minutes. Peel off the paint with a razor blade, then wash the windows with one of the cleaning solutions above.

Appliances

Most household appliances are built well, used a lot, and pick up lots of kitchen dirt and grime. They should be cleaned often with a heavier-than-normal cleaner. Be sure to wear protectove gloves when using the cleaners in this chapter because the acid bases in the recipes can really give you dishpan hands.

Oven Cleaner and Drain Cleaner in One

- ¼ cup (59 mL) baking soda
- ¼ cup white vinegar

Mix into a paste and use a scrub pad to clean oven. The recipe will foam so mix it in a large container. This is basically a 50/50 mixture. Make up just enough for use as it does not really store well. This same formula has long been recommended as a drain cleaner. Just pour it down your drain and flush with hot water.

Stubborn Oven Cleaner

- ¼ cup (59 mL) ammonia
- ¼ cup baking soda
- ¼ cup white vinegar

Preheat your oven to its lowest setting; once light goes off, turn off stove. Put ammonia in a bowl or pan and let sit in the oven for 4-6 hours. Remove ammonia and add vinegar and soda to make a paste. Use the paste and a heavy duty sponge to scrub the stubborn areas. Mist the oven with water and wipe clean. This cleanser also works great for cleaning your barbecue grill.

Citrus Garbage Disposal Cleaner

- 1 lemon
- 1 orange
- Baking soda

Cut strips of lemon and orange peels and place in a bowl. Sprinkle a generous amount of baking soda over the strips. Push several strips down your garbage disposal and turn on disposal and water. The baking soda helps clean the blades while the citrus freshens the air.

All-Purpose Cleaner

- 1 cup (237 mL) liquid soap blend (See page 64 for recipe.)
- ¼ (1.25 mL) teaspoon baking soda
- ¼ teaspoon tea-tree extract
- ⅛ cup (30 mL) isopropyl alcohol

Mix the ingredients together in a spray bottle. This cleaner works great on stove tops and the outside of refrigerators and microwaves. Shelf life: 6-plus months.

Refrigerators

To clean the outside of your refrigerator, use the all-purpose cleaner above. For glass shelves, use the glass cleaner on page 54 and leave an opened box of baking soda in fridge to absorb any odors. Replace every 3-4 months.

Helpful Hint

Is your dishwasher leaving spots on your dishes? Use equal amounts of borax and baking soda and add to your dishwasher soap when doing dishes.

Dishwasher

- 2 teaspoons (10 mL) baking powder
- 1 teaspoon (5 mL) cream of tarter
- 1 teaspoon borax

Add the mixture to the soap holder of your dishwasher when it is empty and run a cycle of hot water. This recipe will clean and sanitize your dishwasher, making it "sparkle" again.

Small Appliances

- ½ cup (118 mL) white vinegar
- ½ cup lemon juice
- ¼ cup (59 mL) water

Blend the liquids together and let them soak on the stain, then scrub. You can put this in a covered container and store in your refrigerator for several months. This formula works great for cleaning coffee and tea pots, cleaning hard water scum, and cleaning grease build up. Use caution if the appliance is colored.

Counter Tops

These fast and easy multipurpose cleansers will clean tile, vinyl, fiberglass and painted woodwork and metal. The first is very gentle, the second is a little stronger, and the third is more commercial and will sanitize the best.

Handy-Dandy Pine Cleaner

- 1 cup (237 mL) liquid soap blend (See page 64.)
- ½ cup (118 mL) pine oil or extract
- 6 cups (1.4 L) warm water

Mix all ingredients together and store in a plastic jug. Spritz on and wipe or wash off. Store in a cool, dry place. Keep out of the reach of children. Shelf life: approximately 6-9 months.

Lemon Fresh Cleaner

- 1 cup liquid soap blend
- ¼ cup fresh squeezed lemon juice
- ¼ eyedropper of tea-tree extract
- 6 cups warm water

Mix together the above and shake well. To use, put in a plastic spray bottle, spritz on, and wipe off. For larger jobs, put in a large bucket, with sponge for large jobs. Spritz on and wipe off, or sponge on rinse sponge then use water to wash off. Store in a cool dark place out of the reach of children. Shelf life: approximately 9-10 months.

Heavy Cleaning Solution

- ½ cup (118 mL) sudsing ammonia
- ½ cup liquid soap blend (See page 64.)
- ¼ cup (59 mL) white vinegar
- 6 cups (1.4 L) of warm water

Mix the ingredients together and store in a plastic jug. Spritz on and wipe or rinse off. Store in a cool, dry cool place. Keep out of the reach of children. Shelf life: approximately 4-6 months.

Fabrics and Fibers

I'm sure Shakespeare's famous phrase—"Out damned spot!"—has been heard on many a cleaning day. This chapter contains cleaning recipes for day-to-day cleaning as well as recipes for pre-treating hard-to-wash spots and spot-treating troublesome stains.

There are four general rules for successfully working with fabrics and fibers. First, try to get to the stain and remove it as soon as possible. Second, always try the easiest and least radical method of removing the stain first. Third, be sure the fabric you are working on is color safe. If you're unsure, blot a damp, white cotton cloth against the fabric in an inconspicuous place. If there's color on your cloth, then the fabric is not color safe. Last, always make sure to rinse the fabric after you do the cleaning procedure.

Laundry Solutions

I don't know about you, but I always thought of laundry day as a real drag. Believe it or not, though, I have found a fun way to do it. I combine a movie marathon afternoon with laundry day. I rent a bunch of movies and only put them on pause long enough to run things back and forth to the washer and dryer. Then I fold while I watch the movies! Between loads I make popcorn.

The following homemade solutions will make your laundry life easier. They work as well as commercially sold products and are much less expensive.

Basic Recipe for Soap Flakes

- Using a cheese grater, grate a bar of gentle castile soap or gentle glycerine soap.*

*Do not use non-foaming varieties.

Basic Recipe for Gentle Liquid Soap Blend/Base

- ¼ cup (59 mL) glycerine soap flakes (See Basic Recipe for Soap Flakes above.)
- ¾ cup (177 mL) boiling water

Mix the soap flakes and the boiling water and stir until the soap flakes have dissolved. Pour into a bottle and label Gentle Liquid Soap Base. Shelf life: 4 to 6 months. If your soap hardens during storage, warm it and add water until the original consistency is achieved.

All-Purpose Washing Machine Laundry Soap

- ½ cup (118 mL) baking soda
- ½ cup powdered castile soap or make ¾ cup (177 mL) unpacked soap flakes*
- ¼ cup (59 mL) washing soda (sodium carbonate)
- ¼ cup borax

Mix the ingredients together and use as you would a commercial washing detergent (about ½ cup per full load). The finer and dryer the soap flakes before mixing with the other ingredients, the longer the shelf life. To prolong the shelf life, use the powdered castile soap or grate your soap flakes very fine and let them air-dry for a day or two before to mixing. Shelf life: 4 months to more than a year, depending on dryness of formula.

Hard Water Laundry Soap

- ½ cup (118 mL) baking soda
- ½ cup powdered castile soap or make soap flakes*
- ½ cup washing soda (sodium carbonate)
- ½ cup borax

To make and use, follow the All-Purpose Laundry Soap mix described above. Shelf life: 4 months to more than a year, depending on dryness of formula.

* The soap flakes must be from glycerine or castile bar soaps; resist the temptation to use anything else.

Pretreating Spray for Regular Washables

- ¼ cup (59 mL) gentle liquid soap base (See recipe on page 64.)
- 1 cup (237 mL) white vinegar
- ½ cup (118 mL) baking soda
- ¼ cup ammonia
- 2 quarts (1.92 L) warm water

Mix the ingredients together until the baking soda is completely dissolved. Store in a gallon jug and pour what you need into a bottle with a spray pump top. Spray this on the dirtiest spots on your laundry before putting in the washer to loosen the dirt and grime. Shelf life: approximately 9 months.

Pretreating Spray for Delicates

- ½ cup (118 mL) gentle liquid soap base (See recipe on page 64.)
- ½ tablespoons (30 mL) ammonia
- 4 tablespoons (60 mL) hydrogen peroxide

Mix the ingredients and test a small sample in an inconspicuous area to make sure the fabric is color stable. Apply a small amount to soiled area and leave it on for 10-20 minutes. Rinse off and wash item. Caution: Do not add chlorine bleach to this mix because it will give off dangerous fumes. Store in a cool, dry area in a glass bottle. Shelf life: 2 to 3 months.

Grandma's Special Handwash Blend

- 2 cups bran (474 mL)
- 1½ quarts (1.44 L) boiling water

Put the bran in a cheese cloth and tie it up like a big sachet. Toss your bran sachet into a pan of boiling water, turn down the heat, cover, and simmer for 30 minutes. Remove from heat, cool, and strain the liquid. Use the liquid for the washing water for fine washables like fine blouses and woolens. This bran solution not only soaks clothes clean but leaves a fine stiffness to the fabric so that spray starch is not needed.

Extra Soft Fabric Softener

- 1 tablespoon (15 mL) of unflavored gelatin
- 1½ cups (355 mL) boiling water

Dissolve the gelatin in the boiling water and add it to the final rinse after washing.

Homemade Fabric Softener

- ½ cup (118 mL) baking soda
- 1 tablespoon (15 mL) arrowroot powder
- 1 tablespoon fine riceflour or cornstarch
- 1-3 drops of your favorite essential oil

Mix the ingredients together and put a few tablespoons inside a small sachet made from a tightly woven fabric. Add a few drops of essential oil, then tie the sachet in a tight, secure knot. (You don't want it coming untied and tumbling out in the dryer.) This softener cuts odors and helps soften clothing, leaving them fresh and fragrant. When the fragrance disappears, empty the sachet, refill, and you're back in business.

Fabric Softener and Deodorant

- ½ cup (118 mL) baking soda

To remove heavy odors in really soiled clothing (pets ordors, garage odors, urine odors, etc.) add ½ cup of baking soda to the rinse cycle of your washer.

Homemade Bleach

- 1 cup (237 mL) hydrogen peroxide
- 3 tablespoons (45 mL) lemon juice
- 15 cups (3.55 L) water

Mix the ingredients together and use as you would a commercial bleach product. This is a safe level bleach that is not too harsh and will fight dinginess in fabrics. (Make sure you are working with fabrics that are labeled as "moderately bleach safe.") Store in a cool dark place. Shelf life: approximately 3-6 months.

Grandma's Natural Spray Starch

- 3 tablespoons (45 mL) and 1 teaspoon (5 mL) cornstarch
- 4 cups (948 mL) warm water
- 1-2 drops optional cologne (use cologne only because it is more water soluble)

Just as my grandmother used cornstarch and water to stiffen doilies for Christmas tree ornaments, cornstarch makes a great natural and economical spray starch. Mix the ingredients together and keep in plastic spray bottle. Shake before using. Shelf life: approximately 6 months.

Out Damned Spot!

Here are some solutions for some of the more troublesome stain offenders for fabrics and carpets.

Blood: Wash thoroughly with cold water—never warm—then make a paste of water and baking soda. Apply the paste to the stain and let sit until totally dry. Use a brush to finish removing the stain. Meat tenderizer also seems to work on blood stains. Make a paste with the tenderizer and water and rub into the stain. Wash out in cold water.

Chocolate: Use the same solution as for Red Wine/Colored Alchohol. (See page 73.)

Coffee: Sponge the stain with a mix of cold water and glycerine.

Egg: Make a soak solution of ¼ cup (59 mL) glycerine, 3 tablespoons (45 mL) baking soda, and ½ cup (118 mL) cold water. Use the solution as a pre-treating spray. Rub it in, then wash.

Fruit Juice: Use the same formula for Red Wine/Colored Alcohol.

Grass: Mix ¼ cup (59 mL) rubbing alcohol, ¼ cup warm water, and ⅛ cup (30 mL) glycerine. Use as a pre-soak and then wash with warm sudsy water.

Gravy: Gravy will usually loosen and wash clean with warm detergent suds, but if the stain is stubborn try rubbing on a mixture of 1 teaspoon (5 mL) ammonia, 1 cup (237 mL) cool water, and ½ teaspoon (2.5 mL) table salt.

Grease and Oil: There are two ways to approach these stains depending on the fabric involved. For delicate fabrics, try very hot, soapy water with a small amount of glycerine added. For more durable fabrics, try spot-treating with a small amount of gasoline or turpentine mixed with glycerine (2 parts gasoline or turpentine to 1 part glycerine), then wash in warm to hot water. For oil and grease stains on durable fabrics that are still "wet," sprinkle salt on the stain to soak up the stain, then wash as described above.

Gum: Use ice to harden, then cut out.

Ink: Mix equal amounts of Cream of Tarter and lemon juice. Let soak for 30 minutes or so, then follow with a cold water wash.

Kid Drinks: Treat fresh stains with a soak in a 50/50 blend of salt and warm water, then wash as usual.

Lipstick/Crayons: If necessary, use ice to harden, then scrape off as much as possible. Follow with a pre-treatmenting rub of 1 teaspoon (5 mL) glycerine, 1 drop lemon essential oil, and 1 teaspoon petroleum jelly. Rub only the affected area, then wash in water as hot as your fabric can tolerate.

Mildew: Newly formed stains will respond to hot, soapy water with a baking powder rinse (see fabric softener rinse above). Old stains have to be harshly treated, so the following recipe is a departure from some of my more natural suggestions. Mix together a solution of ¼ cup (59 mL) chlorine bleach, 1 cup (237 mL) cold water, soak for about 10 minutes. Follow with a 10-minute clarifying soaking in ¼ cup vinegar and 1 cup cold water. Follow with a normal wash.

Mud: Most mud comes out easily enough, but it can be stubborn when ground into fabric. To remove stubbon mud stains, mix together 2 tablespoons (30 mL) of borax, 2 tablespoons of baking soda, and 1 cup (237 mL) hot water. Wash with the above solution, then follow with a "regular" washing.

Nail Polish and Paints: Give careful thought to how much you value the stained item before choosing a method and do a test run in an inconspicuous area to prevent irreversible damage. For durable fabrics, vinegar is the best natural substance to try. Nail polish remover works, too, but with a price: it also removes many fabric dyes. If the item is delicate, the only pre-treating spray that may work to loosen the paint/polish is 3 teaspoons (15 mL) glycerine, 3

teaspoons apple cidar vinegar, and 3 tablespoons (45 mL) hot water.

For hard, non-fabric surfaces, orange or cinnamon essential oils are effective. (These oils may harm varnished surfaces so be cautious.

Perspiration: I know of two good solutions. The first one is a mixture of 1 teaspoon (5 mL) lemon juice, 1 teaspoon vinegar, and 1 cup (237 mL) warm water. Use the mixture as a pretreating spray. Let it sit for 5 minutes, then wash as usual. The second solution is a mixture of 3 aspirin dissolved in 1 cup of warm water. Use it as a pretreating spray. Let it sit for 5 minutes, then wash as usual.

Red Wine/Colored Alcohols: Colored alcohols can be a challenge. Make a paste from 1 table

spoon (15 mL) borax, 1 table-spoon baking soda, and ¼ cup (59 mL) cold water. Lightly scrub the stain and let sit for 10 minutes or so. Rinse with cold water and let dry.

Rust: Wet the affected area with hot water and then apply a paste made from 1 teaspoon (5 mL) sour milk, 1 teaspoon lemon juice, ½ teaspoon (2.5 mL) cream of tarter, and ½ teaspoon baking soda. Rub the paste into the stained area, then rinse. (Do not leave the paste on for more than 5 minutes.)

Tea: Tea stains can be successfully pretreated with a mixture of 1 teaspoon (5 mL) borax, 1 teaspoon glycerine, and ½ cup (118 mL) warm water. Pre-treat for 5 minutes, then wash as usual.

Tar: First scrape off as much of the tar as possible, then rub a mixture of 1 tablespoon (15 mL) butter and 1 tablespoon orange or lemon essential oil into the hardened area to soften and remove the stain. (Water will set the stain). When the tar is totally removed, apply a pretreating spray to the area, then wash in warm water.

Urine: To remove urine stains, make a paste from 1 tablespoon (15 mL) of baking soda, 1 drop of vanilla essential oil, and 2-3 tablespoons (30-45 mL) of warm water. Apply the paste to the area to pre-treat for 15 minutes, then wash. Urine is very acidic so the baking soda will neutralize the acid and the vanilla will cut the smell (pineapple essential is also effective). If the stains are very dark,

you may need to add some bleach to the wash water as well.

Wax: Begin by scraping off as much wax as possible (use an ice cube to harden if needed). Next, place paper over the stain and hold a hot iron above the paper to melt. The porous paper will absorb the wax, but you will need to replace the paper as needed. Use a pre-treating spray prior to washing.

White Wine/Clear Alcohol: White wine or colorless alcohols will usually respond to a water and vinegar solution of 1 part white vinegar to 5 parts water. Scrub until clean, then wash.

Special Treasures

From silver candlesticks to bronzed baby shoes, treasured collectibles present a special set of cleaning challenges. The materials they're made from often require extra care, and the ornate engraved areas that frequently adorn them make great hideouts for dirt and grime.

Purchasing customized cleaners for special treasures can really do some dollar damage to your budget, and because these special cleaners are infrequently used, it's likely that they will dry up or go bad before you need to use them again.

The specialized cleaning recipes that follow can be "whipped up" just when you need them from inexpensive ingredients. For the most part, these cleaners are just as effective as the commercial ones, and they take just minutes to make.

Silver

Good silver is a joy to behold. Always store your clean silver with silica bags to ward off moisture. (You can make your own sachet bags and fill them with the silica used for drying flowers.)

Silver Tarnish and Build-Up Remover
- Baking soda
- Water

Make a gentle abrasive paste of equal amounts of baking soda and water and gently scrub with an old toothbrush. Discard any remaining paste. Rinse the piece under running water and dry thoroughly with a soft buffing cloth. Follow with the Basic Silver Polish.

Basic Silver Polish
- ½ cup (118 mL) liquid castile soap
- ½ cup liquid whiting or liquid blueing
- ½ cup non-sudsing ammonia
- 1 cup (237 mL) hot water

Stir all of the ingredients together and let cool. Place the cooled mixture in a bottle with a cap. Shake well prior to using, then add a small amount to a soft rag or sponge and polish. Rinse the silver with running water and then dry thoroughly with a soft buffing cloth. It is important not to use anything abrasive on your silver because it will scratch. Shelf life: approximately 3-4 months. For silver with build-up in hard-to-reach, intricate surfaces, use the Silver Tarnish and Build-Up Cleaner before you begin to polish.

Food Stain Remover for Silver
- 2 tablespoons table salt
- 1 teaspoon warm water

Mix the ingredients together. Gently rub the mixture into the silver with your fingers, then rinse well. Discard any remaining cleaning mixture.

Gold

Gold is a soft metal, so most contemporary pieces have a gloss over them or are plated with a harder metal for added strength. This gloss can wear off, so gentle cleaning is a must. Older gold pieces made from pure gold also need gentle cleaning care to prevent damage to the soft metal finsih.

Gold Cleaner and Polisher

- ¼ cup (59 mL) liquid castile soap
- 2 cups (474 mL) warm water
- 1 teaspoon (5 mL) ammonia

Mix together and store in glass bottle. Wash the gold with the solution. Dry and polish with soft cloth. Shelf life: approximately 6 months.

Gold Cleaner for Detailed Areas
- Pale beer

Put a small amount of beer on a soft cloth and rub until dry.

Copper

Copper is also a soft metal, although the softness can vary depending on the manufacturer of the piece. The blackish coating that covers copper is easy to remove, but be sure to use a disposable rag because it probably won't come clean again.

Copper Polish

- ½ cup (118 mL) white vinegar
- 1 tablespoon (15 mL) liquid whiting
- 1 tablespoon cornstarch
- ½ cup liquid castile soap
- ½ cup warm water

Use a hand mixer to blend the liquids with the cornstarch, shaking until the cornstarch is distributed. Put some on a soft cloth and polish.

Rinse, dry, and buff with a dry, soft cloth to finish. Shelf life: approximately 1 week.

Copper Cleaner

- Ketchup
- Water

Mix the ketchup and water together in equal amounts. Apply the mixture to the copper with a cloth, then wipe off.

Brass

Brass dulls easily, and brass should shine! These nifty formulas will help bring back the life in your brass.

Brass Polish for Solid Brass

- ½ cup (118 mL) ammonia
- ¼ cup (59 mL) water

Combine the liquids, then dip your brass in the solution. Rinse with water and dry thoroughly. Caution: If you're not sure whether the piece is solid brass, do not use this recipe since it may be too harsh for a brass-coated item. Shelf life: up to 6 months.

Ad Hoc Brass Cleaner

- Worcestershire sauce

Believe it or not, Worcestershire sauce also cleans brass. Just dab some on a cloth, rub on, and wipe off.

Polish for Brass-Coated Items

- ½ cup (118 mL) liquid castile soap
- ¼ cup (59 mL) white vinegar
- ¼ cup liquid whiting

Follow the mixing and cleaning directions in the Solid Brass cleaning recipe. Shelf life: approximately 3-4 weeks.

Pewter

Beautiful pewter needs special handling. The liquid whiting and the alcohol in the recipe below will clean and clarify the pewter, while the ashes, believe it or not, will help make your pewter shine.

Pewter Polish

- ⅛ cup (29 mL) whiting
- 2 tablespoons (30 mL) ethyl alcohol
- Enough fine wood ashes to make a paste (approximately ¼ cup)

Mix together to make paste. Rub the paste onto your pewter with a soft cloth. Rinse and buff dry with clean, dry cloth. Discard any remaining paste by diluting it with running water and letting it go down the drain. The alcohol and the whiting are water soluble; the ashes won't cause drain problems as long as they were fine.

Ivory

I prefer faux ivory because of the inhumane and illegal way ivory is harvested from endangered species, but some people have old ivory pieces in their family treasures and most people don't know how to clean ivory, so here are some tips.

Ivory Polish

- ¼ cup (59 mL) of plain yogurt
- 1 tablespoon (15 mL) lemon juice

Mix the yogurt and lemon juice together. Rub the mixture onto the ivory, let sit for 1-2 minutes, then rinse off. No shelf life.

Ivory Cleaner

- Peroxide
- Water

Mix the peroxide and water together in equal amounts. Rinse well, then follow with the Ivory Polish.

Marble

Marble is very porous and if dirt is deeply imbedded, it may need to be sandblasted. If your once-stunning marble has dulled from dirt and grime, here are two cleaning options to try before resorting to sandblasting.

Light Dirt Marble Cleaner

- Peroxide
- Water

Mix the peroxide and water in a spray bottle. Spray onto the marble, then rinse. Shelf life: approximately 6 months.

Heavy Grime Marble Cleaner

- ¾ cup (177 mL) borax
- ¼ cup (59 mL) water

Mix the ingredients into a paste and use a sponge to scrub. Discard any remaining paste.

Stainless Steel

Stainless steel is a popular kitchen metal because its hard surface is not pourous, making it easy to thoroughly clean. If you take good care of your stainless steel pieces, they will serve you for life.

Baked-On Dirt Remover

- Stainless Steel Polish
- Borax
- Water

Let the Stainless Steel Polish sit on the item for about 1 hour. Rinse well, then follow with a 50-50 mix of borax and water. Scrub with an abrasive sponge and plenty of elbow grease. Rinse well and your troubles are over. No shelf life.

Stainless Steel Polish

- ½ cup (118 mL) ammonia
- ½ cup water

Mix the ingredients in a spray bottle. Spray on and rinse off. Shelf life: approximately 6 months.

Bronze

Bronze is a hard metal that does not respond well to abrasives, but it can be cleaned effectively with oils.

Bronze Cleaner and Polish

- ¼ cup (59 mL) linseed oil
- 2 tablespoons (30 mL) safflower oil
- 2 drops orange essential oil

Mix the oils together and store in a pretty bottle. Shelf life: approximately 6 months.

Computers, Cars, and Other Toys

When we think of our work spaces and our recreational toys, it's hard to think of cleaning. Nevertheless, keyboards and hubcaps do need cleaning from time to time.

One afternoon when I was perusing the auto section in a large hardware store, I found myself in a long row of cleaners, waxes, and the like for autos and boats. I was amazed at the prices!

Although we work hard to find economical cleaning materials for inside the house when we grocery shop, the guys are off buying "liquid gold" cleaning compounds. I made some remark about this and my husband Byron said, "Well, do you think you could do better?" Realizing his mistake, he put the product back on the shelf, smiled, and was gracious when I later brought him an array of cleaning compounds in recycled bottles.

The following recipes are both easy and inexpensive to make, and the results you get may even help you enjoy your toys a little more.

Home Office Equipment

Whether your office is at home, across town, or in your car, here are some helpful solutions to some everyday irritations. The keyboard and computer screen always seem to attract bits of guck and goo, then dust comes along and compounds the problem.

All-Purpose Desk and Equipment Duster

- 6 pieces of soft flannel or cotton cut into 12-inch (30 cm) squares
- 1 cup (237 mL) sweet almond oil
- 2 tablespoons (30 mL) isopropyl alcohol
- 2 tablespoons liquid soap (See recipe on page 64.)
- 1½ cup (355 mL) water

Mix all of the ingredients together and put in a spray bottle. If you're making the recipe as a gift, write the following directions on a decorative tag and tie it around the bottle. "To clean up your office in a jiffy, just shake the bottle several times and make a clean wish. Spray onto your desk and equipment and wipe with these special cleaning clothes and your clean wish will come true." Shelf life: approximately 6 months.

Anti-Static Wipe for Computer or Filter Screens

- 6 pieces of soft flannel or cotton cut into 12-inch (30 cm) squares
- ½ cup (118 mL) commercial liquid fabric softener
- ½ cup water
- 1-2 drops of essential oil if desired

Mix and use as needed in the winter when static shock is everywhere! Can be used on your equipment or on your cloths!!! Just apply some to the fabric square and wipe the static-affected area!

Or you can skip the fabric part and do this formula as a spray-on, just be careful around your computer. If you treat your computer and TV screens with this, dust will not gather so quickly.

Keyboard and Computer Screen Cleaner

- ½ cup (118 mL) isopropyl alcohol
- ½ cup water
- 1 tablespoon (15 mL) baking soda

Mix well and keep near your computer for quick touch ups. Use it as a spray on your screen, but be careful to use a cotton-tip on and around the keyboard so the contacts under the keys don't get wet. Shelf life: 3 to 6 months; can be stored in a spray bottle.

Car and Boat Care

Cleaning "big toys" can be rewarding when the results are dazzling. These cleaning recipes are made from the same types of ingredients as found in expensive, store-bought cleansers

Car and Boat Cleaning Solution

- ½ cup (118 mL) liquid soap (See recipe on page 64.)
- ¼ cup (59 mL) baking soda
- 1 tablespoon (15 mL) alcohol
- 1 gallon (3.84 L) very warm water

Mix the ingredients together in a large bucket. Fill a gallon plastic jug with the mixture, then wash your car with the mixture remaining in the bucket. Shelf life: approximately one year.

This gentle but effective washing solution will clean cars, boats, bicycles, dirt bikes, motorcycles, snowmobiles, and jet skis. The cleanser is mild enough that it does not dull paint or remove wax finishes, yet it cleans toys thoroughly without leaving a residue. My husband says that he likes this solution as a gift when it comes with a gift certificate for one free car wash (labor included).

Tar and Pitch Remover

- Olive oil

Rub a little olive oil onto the offending material. Allow the oil to soften the material, then carefully scrape off.

Winter Build-Up Remover

- Baking soda
- Water

Mix equal amounts of baking soda and water in a bucket. Gently scrub the car and rinse, then wash with the Car and Boat Cleaning Solution above.

Sidewall Tire Cleaner

- Baking soda
- Water

Mix the solution as directed for the Winter Build-Up Remover. Briskly brush the solution onto the sidewalls, then wash with the Car and Boat Cleaning Solution on facing page.

Homemade Car and Boat Wax

- 1 cup (237 mL) carnauba wax
- 1 cup beeswax
- ¼ cup (59 mL) olive oil

Melt the ingredients together over low heat, then pour into a tin and let it set up. Apply as you would any paste wax and buff well.

Leather and Vinyl Upholstery Cleaner

- ⅛ cup (30 mL) liquid soap base (See page 64 for recipe.)
- ⅛ cup baking soda
- ¼ cup (59 mL) alcohol
- 2 cups (474 mL) water

Mix the ingredients together and keep in spray bottle. To clean, spray on a light misting of cleaner and gently wipe off. When working with leather, you may want to use the following moisturizing treatment after cleaning. Shelf life: Shelf life: 4 to 6 months.

Leather Moisturizer

- ¼ cup (59 mL) wheatgerm oil
- ⅛ cup (30 mL) sweet almond oil
- 2 tablespoons (30 mL) castor oil

Mix all of the oils together and apply to leather. Let sit for 1-2 hours, then buff with a dry cloth. This treatment keeps leather upholstery soft and supple and will prevent age cracks. Shelf life: 9 to 12 months.

Windshield De-Icer

- 1 cup (237 mL) propylene glycol
- 2 quarts and 2 cups (2.4 L) water
- 4 cups (948 mL) isopropyl alcohol

Mix together and keep in a spray bottle and spray on as needed. This formula is inexpensive to make and contains the same ingredients as commercially sold de-icers. If you cannot find propylene glycol, then substitute the same amount of a commercial antifreeze (most commercial antifreezes are made of from propylene glycol.) Keep out of reach of children and pets. Shelf life: 7 to 12 months.

Rust Deterrent

- 2½ cups (592 mL) raw linseed oil
- ½ cup (118 mL) turpentine

Mix the materials in an amber glass container and label with the contents. Paint the solution onto the item you're rust-proofing with a paint brush. Let sit for several hours, then wipe off any excess liquid with a dry rag. Always wear rubber gloves and protect your working surfaces when working with this solution. Apply the above solution to any metal to help prevent rust. It's great for cars on the coast, and even for gardening tools in rainy climates.

Battery Cleaner

- 1 cup (237 mL) baking soda
- 1 cup water

This cleaner neutralizes the battery corrosion on posts and cables that can cause car start-up problems. Paint the paste on with an old paintbrush. Let sit for one hour, then clean off. Dispose of any leftover cleanser by pouring it down the drain with running water.

Homemade Antifreeze for Gas Tanks

- 2 cups (474 mL) isopropyl alcohol
- 2-4 drops fir essential oil or cinnamon essential oil

Mix the ingredients and keep in glass container. Add ¼-½ cup (59-118 mL) to your fuel tank every other week in colder climates. The alcohol mixture acts as a solvent and helps remove water that can get in your gas tank in the form of ice, and also helps to prevent the gas from freezing. The essential oil will accelerate this process. Shelf life: 6 to 9 months.

Starting Fluid

- Ethyl alcohol

Most starting fluids are straight ethyl alcohol (grain alcohol). To make your own, just put some in a glass spritzer. When your car won't start, spray a little into the air intake valve on the carburetor and start the car. (Refer to your owner's manual if you are not absolutely certain of the location of your carburetor's air intake valve.) The process can be repeated only once if needed. Shelf life: 12-plus months.

Outdoor Areas

For many people, the outdoor areas of their home bring every bit as much satisfaction and enjoyment as the indoors. I love to work outside in my garden. Where I live, the spring/summer season is short and I spend as much of that season outside as possible. (I have even talked my husband Byron into building a greenhouse attached to our house so I can extend my gardening pleasure as long as possible.) This chapter contains recipes for tackling a range of outdoor cleaning challenges—from patios to siding—with inexpensive, natural ingredients.

Outdoor Home Cleaning

Cleaning the areas outside of your main living area can be just as rewarding as indoor cleaning. A freshly cleaned deck or walkway can reawaken pride and love for your home.

Moss Cleaner for Decks
- Apple cider vinegar

Pour undiluted apple cider vinegar over the deck. Let sit for 1 hour, then rinse.

Multipurpose Garage Cleaner
- 1 gallon (3.79 L) water
- 1½ cup (355 mL) baking soda
- ½ cup (118 mL) vinegar
- 3 tablespoons (15 mL) alcohol

To clean up most tools and keep paint brushes fresh, give them a quick cleaning with this solution. To help loosen a recent oil leak, pour some of the above solution on it and cover with pure baking soda. Let sit overnight and sweep up in the morning. To make the recipe, combine the ingredients and shake until mixed.

Cleaning Wood or Vinyl Siding
- 1 gallon (3.79 L) water
- 4 cups (948 mL) chlorine bleach
- ⅔ cup (158 mL) trisodium phosphate
- ½ cup (118 mL) borax

This recipe will restore a fresh look to your siding and will revitalize the look of the color. To make the recipe, just mix the ingredients together. Scrub on with a brush and rinse off. Keep out of the reach of children and pets, and protect your hands with gloves. Discard any unused material.

Pool and Spa Cleaning

- ½ cup liquid soap blend
  (See page 64.)
- 1 teaspoon borax
- 2 teaspoons baking soda
- 1¾ cups very warm water

Pour the water into a stainless steel mixing bowl and add the liquid soap. Stir well, then add the dry ingredients. Stir the mixture until it's grainy. Store in a squeeze top container (old ketchup containers work well) and stir or shake before using. Use with scrub brush. Keep out of the reach of children. Keep the cleaner in a cool, dry place.

Shelf life: approximately 3-4 months. Note: If you have a stubborn stain on a fiberglass or plastic surface, you can add a few drops of peroxide to this mixture and "scrub" away the stain.

Helpful Hint

If your pool or spa pH is low(less than 7.5) you can raise it by adding small amounts of baking soda (sodium bicarbonate) until the pH rises to where it should be. If your pH is too high (7.8 or above, which is common in hard water areas), then lower your pH by adding sodium bisulfate a little at a time.

Stain Remover for Concrete Walks, Patios, & Floors

- 1 cup (237 mL) alcohol
- 2 ounces (60 mL) essential oil of cinnamon or oil of fir

Combine the ingredients and stir well. Wash thoroughly with a basic soap and water compound and rinse well. Apply to the problem spots and let sit for 2 - 4 hours. Rinse well. Keep pets away as they will want to lick the mixture and it is caustic.

Cleaning Outdoor Tile

- 2 cups (474 mL) very warm water
- 4 tablespoons (60 mL) baking soda

Dissolve the baking soda in the warm water. Prepare the area by washing it thoroughly with a basic soap and water compound and then rinsing. Scrub the area with the baking soda recipe and let it sit for 2-4 hours. Rinse thoroughly.

Cleaning Stubborn Outdoor Tile & Plaster

- 2 cups (474 mL) water
- 4 tablespoons (60 mL) muriatic acid

For more stubborn problems such as unwanted mortar on tile and plaster stains, here is a more commercial solution. While wearing protective gloves, add the acid to the water (never vice versa). Apply to the trouble area and let sit for not more than an hour. Muriatic acid is caustic, though, so rinse thoroughly and keep children and pets away.

Pets

Pets bring so many tender moments and smiles to our daily lives that most of us can't imagine our lives without them, but they also add a few extra cleaning chores to our already-full slates. Removing odors, keeping fur clean, and fighting ticks and fleas are frequent pet-related cleaning projects.

Pampering our pets with natural cleaning products is a great way to thank them for their quiet, loyal love. I developed these recipes after spending some time wandering the pet sections in stores to determine the most common pet care challenges. The recipes are simple and inexpensive to make, and you can control the type and amount of fragrance in the finished products.

Odor Fighters

As much as we love our pets, we hate their odors. The delight of a new, cuddly puppy often vanishes when we're confronted with fresh puddles of urine soaking into our living room carpet. There are two ways to deal with pet odors. One is to cover them up; the other is to neutralize the smell. The following recipes will help you neutralize and minimize these odor problems.

Wood and Vinyl Floor Odor Remover

- 1 cup (237 mL) vinegar
- ½ cup (118 mL) baking soda
- 4 cups (948 mL) water

Mix ingredients together and scrub well. Shelf life: approximately 3 months.

Carpet Odor Remover

- 1 cup (237 mL) vinegar
- ½ cup (118 mL) baking soda
- 4 cups (948 mL) water
- Hyperdermic needle
- Vanilla and pineapple essential oils

Mix the vinegar, baking soda, and water together and scrub the carpet well with it. Allow to thoroughly dry. Mix the two essential oils together in a 50/50 ratio, then inject small amounts of the oil mixture through the carpet and into the carpet padding with a hypodermic needle to "neutralize" the odors.

Litter Box Deodorizer

- Baking soda

Baking soda is the simple answer here—just mix a little with the kitty litter. Do not be tempted to overpower the smell with essential oils because the fragrance can sometimes deter the cat too much and they will go behind the couch!

Bathing

I have two large dogs and one small one, and on bathing day my husband often remarks that it's difficult to tell who had the bath! When using the following recipes, be sure to keep them away from your animal's eyes. Dogs are sensitive to smell, so try not to over perfume them to mask odors. The dry shampoo is great for cats (who generally don't like water) and for dogs who need touch-ups between regular baths.

Help! My Dog Played with a Skunk!

- 1 box baking soda
- Juice from one lemon
- 1 cup (237 mL) vinegar
- 6 drops of vanilla essential oil
- Small amount of shampoo

Mix the ingredients in a large jar or bowl, leaving lots of room for expansion when it foams up. Pour a tub of warm water and add the mixture. Repeat if needed.

Gentle Flea and Tick Repellent Bath

- 1 cup (237 mL) gentle liquid castile soap
- 2 cups (474 mL) warm water
- 3 drops each of citronella, thyme, rosemary, and cedarwood oil

Mix all together and shake before using. Shampoo animal twice and rinse. Keep out of eyes. Shelf life: approximately 6 months.

Dry Shampoo (deodorizing and flea and tick repellent)

- ½ cup (118 mL) arrowroot powder
- ½ cup corn starch
- ½ cup baking soda
- 3 drops each of citronella, thyme, rosemary, and cedarwood oils

Sift the ingredients together into a powdered mixture. You may need to sift a few times to force the essential oils into to the powder. Working outdoors, sprinkle the powder on the animal and then brush off. This is great for between shampoo touch-up for dogs and great for cats who do not like water. Package in a container with sifter top (a old "cheese or herb" shaker will do).

Flea and Tick Fighters

Fleas and ticks are nobody's friend. They make our pets' lives miserable and also carry diseases. After you've rid your home of fleas and ticks, the best thing you can do is to take the natural approach to repel them. Cedar beds are one way to discourage them, and so are the natural compounds in the following recipes.

Flea and Tick Compound

- 12 drops rosemary
- 12 drops cedarwood
- 12 drops thyme
- 12 drops citronella

Mix together the following essential oils to make a "compound" that will be used in the next few items you can make for your pet. Store in an amber or dark-colored bottle. Shelf life: 6 to 8 mnths.

Flea and Tick Spray

- 1 tablespoon (15 mL) liquid soap (See recipe on page 64.)
- 1 teaspoon isopropyl alcohol
- ½ cup water
- 6 drops of Flea and Tick Compound

Mix the ingredients together and put in a colored glass spritzer bottle. Shake before using. Spray on pets and in areas that pets frequent to discourage pests.

Helpful Hint

Make a pet bed out of cedar chips to discourage fleas from the pet and sleeping area. Spray the perimeter of yard with the Flea and Tick Spray described above. Add drops of essential cedar oil of cedar from time to time to refresh.

Flea and Tick Powder

- ½ cup (118 mL) cornstarch
- ½ cup arrowroot powder
- 12 drops of Flea and Tick Compound on facing page

Sift several times to force the essential oils into the powder. Put in a decorative shaker container and sprinkle on when needed. Store in an airtight container. Shelf life: 6 to 8 months.

Flea and Tick Collars

Make the flea and tick powder from the recipe above. Measure you pet's neck and add 5 inches (12½ cm) to the measurement. Cut a cotton fabric strip measuring approximately ½ to 1 inch (1¼ to 2½ cm) ide plus the length calculated from above. Seam together lengthwise to make a long tube and turn right sides out. Measure in 2½ inches (6 cm) from one end (this leaves one side of the tie) and make a seam there to seal the tube. Stuff the tube with the flea and tick powder, stopping 2½ inches from the top, and sew a seam to close the tube. Use the empty parts of the tube on each end as ties and tie on your pet's new natural flea and tic repellent collar.

(Note: If your pet spends time outdoors, add 1-2 inches to the length measurement. Stretch and sew some elastic to the wrong side of the fabric in the middle of the strip's length. Finish the collar as directed above. Adding elastic ensures that your pet can wiggle out of the collar if they get caught up in something around their neck.)

Great Gifts

Most of the recipes in this book can be arranged in a beautiful package for a great gift. "What?" you say, "Give cleansers as a gift? Give me a break!" While giving drain cleaner by itself as a gift may not be appealing, it can easily be part of a special gift when packaged in a thoughtful container with a few other small gifts tied to the same theme. The thoughtful gift ideas below will dazzle you (and their recipients!) with both their beauty and their practicality.

Handyman's Gift Basket

- Plastic garbage pail in a bright color
- Concrete Stain Remover
  (See page 98 for recipe.)
- Multipurpose Garage Friend
  (See page 96 for recipe.)
- Fireplace, Brick, and Stone Cleaner
  (See page 47 for recipe.)
- Decorative rags
- Putty knife
- Squeegee
- Length of raffia

This basket makes a great gift for all of the special men in your life. Arrange all of the items in the bucket, then tie the raffia around the bucket.

New Job or Office Gift

- Basket
- Spray bottle of Anti-Static Wipe for Computer or Filter Screens (See page 86 for recipe.)
- Spray bottle of Keyboard and Computer Screen Cleaner (See page 87 for recipe.)
- Bottle of All-Purpose Desk and Equipment Duster (See page 86 for recipe.)
- Decorative cleaning rags
- Coffee mug
- Mug warmer
- Gourmet coffee
- Picture frame
- Raffia or decorative ribbon

Here in this pretty gift basket is everything a budding executive needs to start off right. Arrange all of the items in the basket, then tie the ribbon into a bow with streamers.

Gardener's Delight

- Large garden pot
- Garden tool, seed packet(s), garden gloves
- Bottle of Moss Cleaner for Decks & Planters (See page 96 for recipe.)
- Bottle of Happy Leaves Houseplant Polish (See page 47 for recipe.)
- Container of Stain Remover for Concrete Paths, Patios, & Birdbaths (See page 98 for recipe.)
- Decorative ribbon

Arrange the gift items inside the garden pot, then tie with ribbon. This collection is the perfect gift choice for that special gardener in your life. Use it to celebrate May Day, a birthday, or another special occasion. I'm sure they will thank you with abundance from their garden!

New Car Gift

- Bottle of Car and Boat Cleaning Solution (See page 88 for recipe.)
- Bottle of Leather Moisturizer (See page 90 for recipe.)
- Small putty knife
- Bottle of Windshield De-icer (See page 91 for recipe.)
- Tin of Homemade Car Wax (See page 89 for recipe.)
- Do-it-yourself auto repair book
- Length of raffia ribbon

We used a colorful cleaning bucket again as the "basket" and filled it with an assortment of great cleaning items into it. Arrange all of the items in the bucket, then tie the raffia around the bucket.

Housewarming Gift

- A copy of this book
- Basket
- Bottle of furniture polish (See pages 38-41 for recipes.)
- Bottle of Lemon Fresh Cleanser (See page 60 for recipe.)
- Several fun household sponges
- Feather duster
- Small potted plant
- Happy Leaves Houseplant Polish (See page 47 for recipe.)
- Small packet of tea or coffee
- Decorative bow or ribbon

Many formulas in this book would make nice additions in a housewarming gift basket. Here are the ones I selected for my kids. (Moms of the world unite: with this gift and a little encouragement, they may keep their first apartment cleaner than their "home" rooms!)

New Pet Gift Basket

- Basket
- Dog leash
- Puppy toy
- Dry Shampoo (See recipe on page 103.)
- Gentle Flea and Tick Repellent Bath (See recipe on page 103.)
- Recipe or index card and envelope
- Large bow

Imagine that your friend, son, or daughter has just brought home an adorable, wiggly puppy to bless their home. Here is a cute gift basket idea to let them know you care and to welcome that new little arrival.

Arrange all of the items in the basket. Copy the recipe for removing pet odors from wood and vinyl floors from page 102 onto the card, tuck it in the envelope, and arrange it in the basket. Decorate the basket with a large bow or festive ribbon.

Index